SCIENCE MUSEUM

John Hopkinson
Electrical Engineer

by

JAMES GREIG, M. SC., PH. D., F. I. E. E., F. R. S. E.

WILLIAM SIEMENS PROFESSOR OF ELECTRICAL ENGINEERING
IN KING'S COLLEGE, UNIVERSITY OF LONDON

GW00707874

LONDON
HER MAJESTY'S STATIONERY OFFICE
1970

1   John Hopkinson

Introduction

Amongst the pioneers of electrical engineering, John Hopkinson stands out as an engineer-scientist. He recognised competence in mathematics and physics as an essential pre-requisite to the profession of engineering. He forms an interesting subject for biographical study, for, not only was he distinguished in his career, but he possessed qualities of character which declare him an individualist. In addition, his life is remarkably well documented. The Hopkinson family 'rather ran to memoirs' and a substantial amount of unpublished material has been preserved.

This booklet outlines Hopkinson's contributions to engineering and to pure science and attempts to present something of the character of the man.

I owe a debt of gratitude for much help in the preparation of this little book. My interest in Hopkinson arose through undertaking to give, in 1949, a lecture in celebration of the centenary of his birth. This led me to make contact with some of his relatives, with several of his students and one of his contemporaries. At that time Hopkinson's only surviving child, Lady Ewing, was living in Cambridge. She, from intimate personal knowledge, created for me some picture of the man within his family circle. Two of Hopkinson's students then surviving, F. Lydall and B. Welbourne, were able to speak of his supervision in experimental work. Sir James Swinburne, somewhat younger than Hopkinson but effectively his contemporary, with whom he was closely associated in his legal work, wrote a shrewd critical assessment of him. More recently, Captain T.G.N. Hilken furnished much useful information about Hopkinson's association with the University of Cambridge.

I would acknowledge particularly Mr C. Mackechnie Jarvis's drawing to my attention the existence of unpublished material in the care of Lady Chorley and especially the kindness of Lady Chorley, John Hopkinson's niece, in inviting me to consult and later to borrow all the relevant material in her possession.

Professor Allan Martin made an extensive search of the Edison Archives in Orange, New Jersey. Mr J.W. Wellwood and Mr J. Allcock furnished advice relating to patent litigation. Mr J.E. Wright, Librarian to the Institution of Electrical Engineers, was unfailingly helpful in the provision of references.

Special thanks are due to Mr J.L. Willis, Director, and Mr H.H.G. McKern, Deputy Director, of the Museum of Applied Arts and Sciences, Sydney, who arranged for the arc lighting equipment from the Macquarie lighthouse to be photographed.

Miss D. Rollinson, my secretary, not only typed the manuscript, but contributed substantially to the preparation of material for the text and illustrations. My colleague, Mr Geoffrey Platt, gave valuable help with material relating to electric traction. Miss Margaret Weston, Mr Brian Bowers and Mr John Smart, of the Science Museum staff, have given invaluable guidance and help in the whole preparation of the booklet.

To my wife, whose interest and enthusiastic help have been maintained throughout the twenty years in which Hopkinson has been for me a pleasantly intriguing spare-time study, I express my warm thanks. *James Greig*

Early life and education

The *Edinburgh Review* of 1849, the year in which John Hopkinson was born, contains an article dealing with electricity and its application to the telegraph. Evidently by that time the subject was deemed to merit the attention of the well-informed, the influential and the leisured classes.

It was in June of that year that Michael Faraday communicated to the Royal Society James Prescott Joule's memoir *On the Mechanical Equivalent of Heat* which was to be followed by Kelvin's recognition of the true basis of the thermodynamic cycle. The Great Exhibition of 1851, which in a very real sense was to mark the opening of the technological era, had been proposed and its plan outlined.

John Hopkinson's parents were both of North Country stock, his mother belonging to the Yorkshire cotton spinning family of Dewhurst, his father, an engineer and millwright by profession, being a Lancashire man. Starting his engineering career as an apprentice with the firm of Wren and Bennett in Manchester, John Hopkinson's father had, by the date of John's birth, 27 July 1849, become a partner in the firm and it was in conditions of comparatively solid material comfort that the boy grew up. Sensible though they were of the social and political movements of their time, the parents possessed that stability of outlook, integrity and sense of purpose that characterised, at its best, the

2   John Hopkinson (left), his brother Alfred and sister Ellen

Victorian non-conformist family. For John and Alice Hopkinson, religion and their church – they were Congregationalists – occupied a central and supremely important place in their lives. Competent, resourceful and enterprising in the practice of his profession, a Liberal in politics, active in social reform particularly in relation to the conditions of workpeople, and interested generally in civic affairs, John Hopkinson first appeared in public office in 1861 when he was returned unopposed to represent St Lukes Ward on the Manchester City Council. In 1872 he became an Alderman and in 1882 was elected Mayor of the City of Manchester. That John Hopkinson and his wife were people of independent judgment, prepared to act in accordance with their convictions, was evidenced in the education of John, who was the eldest son. The boy was sent, after a period at a day school near his home, to Lindow Grove in Cheshire. The headmaster of this school, Charles Willimore, was a man of very remarkable qualities, who influenced deeply John's character and subsequent development. In a letter written a few years later to Mr Hopkinson he remarked 'I greatly object to long hours of study for boys. I have a great belief in the physical development being the main thing until a boy leaves school, i.e. for what I have said might be misinterpreted – that any intellectual advance at the expense of health is dearly bought.'

John was not, as a small boy, very robust but at this school, which was situated in delightful countryside, he developed under Willimore's sympathetic guidance a love of the out-of-doors which was to become one of the ruling passions of his life. In 1864, Willimore was appointed to the headmastership of Queenwood School in Hampshire and young Hopkinson went there with him. It was at this stage that the boy's ability for mathematics first showed itself. A letter written from Lindow Grove by John to his father on 2 December 1863, gives a little local colour to the scene when the move to Queenwood was about to take place. Starting, as might be expected, with 'thanks for the money', the letter goes on:

> Next half, Edwin Thompson is leaving and all but certain is going to Owens College. There is another boy from here going there, Ian Holdsworth, he is considerably below me. I should not like to stop here next half because I should be almost top and we are certain not to get as good a master as Mr Willimore. If I go to Queenwood, very probably I should be near the top there, for lately that school has been going down because the last master left about half a year ago. But you can get all you want about Queenwood from Mr Willimore when you see him.
>
> As Edwin Thompson is going to Owens College, I should like to go there but it is very expensive, I think, but I should waste no time there on humbuggy classes through disorder and I could go with all my force into the things I most need. Here in mathematics about 4 hours a week is taken up and in Euclid I can always prepare my lessons in class. Here we have no science, none but botany, even for our own time; there I could do a great deal of science. I have just begun 6 book of Euclid.

There was, in fact, some science teaching at Queenwood and John obtained a fair grounding in chemistry.

It was natural, and evidently the father's intention, that the eldest son should follow his father in an engineering career. The engineer of the 1860s began his apprenticeship in the shops at the age of 14. Experience came first in the scale of values and science was regarded as an incidental rather than an essential ingredient of the engineer's make-up. A new pattern of engineering education was, however, being evolved.

University courses in engineering had been in operation for a number of years in Glasgow and in London; in Manchester, the Owens College – founded some fifteen years earlier – was providing a course of theoretical and experimental study in science. It was in 1868 that the Engineering School was founded there by the appointment of Osborne Reynolds to a Chair of Engineering. John had himself formed the opinion that before taking up engineering he should have some scientific education and with characteristic exercise of independent judgment, Mr Hopkinson acceded to John's wish and sent him in 1864, at the age of fifteen, to Owens College to pursue the three-year course. The college could not, at that stage, award its own degrees but the major courses prepared students for the External Examinations of the University of London. Owens was already building up a high reputation as the result of the labours of a group of able and distinguished teachers. In the list of staff for the session 1864–65 appear the names of H.L. Roscoe, the chemist, and W.S. Jevons, who was the first to popularise the mathematical methods of Boole in the development of symbolic logic.

J.J. Thompson in his *Recollections and Reflections* records that when, some five years later, his own entry to Owens was being considered, a friend remarked to his father 'If I were you, instead of leaving the boy at school, I should send him while he is waiting to the Owens College: it must be a pretty good kind of place, for young John Hopkinson, who has just come out Senior Wrangler at Cambridge, was educated there.'

At Owens, young Hopkinson came under the influence of the mathematicians, Sandeman and Barker, both Senior Wranglers, and almost immediately his exceptional powers began to show themselves in consistent examination successes. At the end of the session he was awarded the Mathematics Prize for the Junior Class (Higher Division). To Barker, Hopkinson owed the suggestion that he might go to Cambridge and read for the Mathematical Tripos. In May 1867 he was admitted as a pensioner in Trinity College and a year later he became a Scholar. Again, the father's independence of outlook had enabled him to approve a course of action for which there could be little precedent at the time – preparation for an engineering career by reading mathematics at the University.

The Cambridge system and young Hopkinson proved to be admirably adapted to one another. He went up determined to work, and if possible to take first place. Rowing and running were his sports and he took his full share in undergraduate life, even in the occasional escapade. He began to show some of the characteristics which, in more mature form, found expression in his later life. One of these was his dislike of the conventional, a healthy scepticism of the accepted and a somewhat puckish delight in iconoclasm. In his second year he became Captain of the Second Trinity Boat Club.

A letter written from Trinity in February 1870 to Evelyn Oldenbourg, to whom he became engaged about two years later, illustrates his undergraduate outlook:

I have just been reading a story of Thackeray's *The Fatal Boots*; it is an auto-biography of an utterly mean scamp who is always trying to swindle and only succeeds in over-reaching himself, and yet he manages to make one identify oneself with this disagreeable hero. Thackeray has a wonderful power of setting forth mean characters. *Vanity Fair* is full of them.

As President of the Athletic Club I put up a notice of meeting in my room this afternoon at five, intending to propose Glaisher as Secretary of the Club; no-one but Glaisher came, so we unanimously elected him Secretary and carried one or two other motions, everything in a most formal manner; fortunately there is nothing in the rules about a quorum and the meeting was duly called so everything was quite valid. Fancy me getting up and addressing Glaisher as 'Gentlemen' and proposing 'that Mr Glaisher be Secretary of the Club', and in flowery language dilating on his merits; of course, he had to second himself and we carried it between us . . .

Boating is going in full swing; the eight goes down at 2.30 and we get back to our rooms in time to dress for Hall at 4.30. At present, I go down every day at 2 to coach men in a pair-oared boat until 2.30. It is rather disagreeable work as one or two are awful duffers and it is perfectly ghastly to see them. I tell them so and it is cold work steering and lecturing on rowing in thin clothing and a cold east wind. I am rowing seven in the eight, that is, next behind stroke.

. . . They have put me on again to read in Chapel, Kirkpatrick, who makes a point of keeping morning Chapel, reads the morning lessons for me and Glaisher will read the Thursday and Friday evening ones, so I shall not have to keep more than the usual three. As I had only the responses of the Latin grace to read, I did not succeed in getting into another row about that.

The last paragraph relates to the constant battle which he waged with the College authorities, trying to maintain his non-conformist independence.

In mathematics, he coached with Routh, the most successful mathematical tutor of the day, and during his course he sat several examinations not connected with the University of Cambridge. In 1868 he passed the B.Sc. examination of the University of London. In the same year he sat for and won a Whitworth Exhibition. The award had just been founded 'for the promotion of Engineering and Mechanical Industry in this country'.

The degree of Doctor of Science had been instituted by the University of London in 1862 and in the form in which it was established was awarded on the result of written examinations only. In 1870, the year before he sat for the Cambridge Tripos, Hopkinson took the examinations for the degree of D.Sc. in the University of London. The degree was not conferred until 1871 and in June of that year his name appeared in the Tripos List as Senior Wrangler. Shortly afterwards he took first place in the Smiths Prize. It is recorded that three weeks before the Tripos examination, he ran the mile race at the Sports, breaking the previous University record.

In the autumn he was elected to a fellowship at Trinity. He was one of the first non-conformists to be elected under new statutes in which religious tests had been abolished but which still required celibacy. With such a record he was already a marked man.

7

Lighthouse engineering

On coming down from Cambridge, Hopkinson joined his father's firm to gain practical experience. Within less than a year, however, Mr Hopkinson was approached by Mr James Chance, head of the Birmingham firm of optical glass manufacturers, with a proposal that John should take up a post with that firm. It was arranged that, after a few further months of experience with his father, he should take up the appointment of engineer and manager of the lighthouse department of Chance's optical works in Birmingham. This he did in March 1872 at the age of twenty-two.

It was at this time that John became engaged to Evelyn Oldenbourg, a girl whom he first met when still an undergraduate at Cambridge, and a school friend of his sister, Ellen.

Although Hopkinson never returned to Manchester to live, he maintained throughout his life close contact with his family and with his native city. As has been mentioned, his father became a member of the City Council in 1861 and was elected Mayor of the City in 1882. One of his brothers was called to the Bar, became a member of Parliament and, later, Principal of Owens College and Vice-Chancellor of the University. Two others became engineers and with one of them, Edward, John was to be closely associated in some of the episodes of his professional life.

On 4 March 1873, John Hopkinson and Evelyn Oldenbourg married. They set up house on the outskirts of Birmingham some five miles from the Smethwick works of Chance Brothers. It is recorded that he rode back and forth to the works on their cob *Pegasus* and the horse was also used in harness in their little Stanhope phaeton. Two rooms in the house were allocated as laboratories for experimental research. One was devoted to magnetism and electricity and the other to optical measurements on titanium glass which apparently was produced in the optical department of the works to meet a special request from Professor Stokes of Cambridge.

It appears that Hopkinson made some of his experiments on the residual charge of the Leyden jar in these 'domestic laboratories', for Mrs Hopkinson wrote: '. . . I remember that it meant a tapping of the jars throughout the night and no doubt throughout the day, every two hours. I heard the tapping by him on the jar as it might be a bird on the window sill all night long'.

It may seem a little strange that the engineering of lighthouses should have directed Hopkinson into the path which was to lead him to be one of the greatest electrical engineers of his day but the reason lay in the application of the arc light as an illuminant for lighthouses. In the five years which he spent with Chance, he made himself an expert in the design of lighthouse optical equipment, was responsible for several important projects and he effected certain advances in practice.

At the time of the 1851 Exhibition the idea that the electric arc might be used as a light source of high intensity for lighthouses was being widely canvassed and this stimulated the design of magneto-electric machines of substantial size. Following some important advances in Belgium and in France, Professor

Holmes of Newcastle designed a machine specifically for lighthouse purposes which proved to be highly successful. An experimental installation made by Holmes at Blackwall in 1857 was assessed by Faraday on behalf of Trinity House and approved, and the following year equipment was installed permanently at the South Foreland lighthouse. By 1880, ten lighthouses were equipped with electric arc lights – five in Britain, three in France, one at Odessa and one at Port Said.

Hopkinson's most noteworthy development was that of group flashing lights. This was a modification of the optical system in which, by arranging the axes of the lenses to form groups, two, three or four flashes at short intervals could be obtained, separated by longer intervals of darkness.

The first important work upon which he was engaged was the Lizard light. Two of the projects with which he was concerned resulted in visits to Europe, the first to Russia and the second to Italy. Mrs Hopkinson took part in both and gives a most entertaining account of them. They sailed from Grimsby to Reval, now Tallinn, in Estonia, where the lighthouse was to be built and went thence by train to St Petersburg.

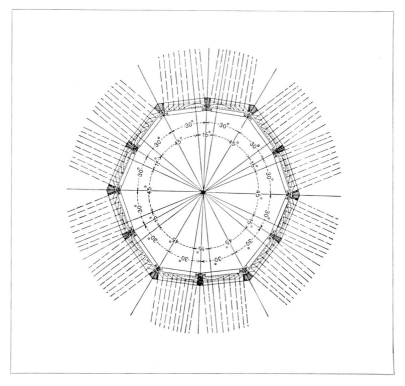

3   The arrangement of lighthouse lens panels to produce the group flashing feature

Mrs Hopkinson was critical of the official incompetence and laziness which her husband encountered, of the miserable condition of workpeople, and of the primitiveness of much in everyday life. Counting in the shops was by abacus. Describing a visit to Tsarskoye Selo and Peterhof, she wrote: 'The Crown Prince and the aristocracy of Russia were to be seen driving in magnificent slender phaetons drawn by fresh and nearly wild, almost uncontrollable, stallions harnessed only in narrow straps covered with innumerable silver rings. The driving was rapid and superb, a masterly display on the terrace below the palace, below which flowed the shining Neva.' Admiral Briganoff sent a naval vessel to take them from Reval to Helsingfors.

The second visit was to the Tino lighthouse located to illuminate a dangerous area between Genoa and Spezia. Mrs Hopkinson recalls that, when she joined her husband at Tino there had been a bad outbreak of cholera at Spezia.

Hopkinson improved the performance of Chance's lanterns by the introduction of agate bearings. He published in 1874 a pamphlet entitled *Group Flashing Lights*. It appears that the company was unwilling to patent this invention and this, in some measure, influenced Hopkinson towards leaving the company's service, which he did in 1877, and setting up as a Consultant. He did not, however, completely sever his connection but remained Consulting Engineer to the company for three further years.

Hopkinson's move to London involved an almost complete change in his professional activities and responsibilities. As engineer, and manager, of Chance's Optical Works he had been concerned with the day-to-day administrative problems of a works department, with routine engineering design and production, and with the specialist design techniques of lighthouse optical systems. He found the management of men less to his taste than the tackling of technical problems. Nonetheless, he was successful in introducing improved production methods based, in some instances, on piecework and, overall, in increasing the profits of his department.

Illustrative of the occasional difficulty with workpeople is a remark in one of his letters to his fiancée during his first year with Chance:

I had a complaint that the boys employed in grinding did mischief in the works during meal hours. I think I have made short work of them by making all responsible for each. I instituted good conduct premiums some little time since, 10/– each at the end of six months; I told them I should stop it for all if I had complaints of any of them and that they must look after their offenders. I expect it will work.

Short scientific papers by Hopkinson began to appear within about a year of his coming down from Cambridge. Two on *The Rupture of Iron Wire by a Blow* appeared in the Proceedings of the Manchester Literary and Philosophical Society. Two others were contributed to the Messenger of Mathematics, one being *On the stresses produced in an elastic disc by rapid rotation*. In the latter paper he remarks ' . . . This problem has a certain practical value. It not infrequently happens that the grindstones used for polishing metal work are ruptured by the tensions caused by rapid rotation . . . '.

It was the coding of lighthouse flashes which first brought Hopkinson into contact with Lord Kelvin, then Sir William Thomson. In Section G, the Engineering Section of the British Association, at the Brighton meeting in 1872, Hopkinson criticised a proposal made by Kelvin. The result was an invitation to the young man 'to join him in his yacht, the *Lalla Rookh*, and sail in it to the Irish Coast and make investigations on the point in dispute'. From this there developed a lifelong friendship and it is clear that Hopkinson enjoyed and valued very highly the opportunity of the exchange of scientific opinion with this great man. Thomson's regard for the young man is evidenced by a letter written in 1878, which it may be worthwhile to quote in full:

Peterhouse Lodge,
Cambridge.
Feb. 3rd, 1878.

Dear Hopkinson,
Your letter has been forwarded to me here, but not the specifications which no doubt I shall find in Glasgow on my return. What would you think of a joint publication to the United Service Association on Distinguishing Lights? I should think the Council would be glad to accept any such proposal. I am to give a lecture there tomorrow evening, 8.30 o'clock, on 'Compass and Sounding Machine' for which I enclose an invitation. If you chance to be free and are inclined to come I shall be glad to see you. We might talk over the Lighthouse question and if you approve make a proposal regarding it to the Secretary. I had often thought of distinguishing different directions by different groups of eclipses or flashes and I rather think I proposed it in my old Good Words article but cannot recollect for certain. I have not seen an easy mechanical way of carrying it out and so have not urged the idea. Do you see it? If either the person you refer to (but don't name – is he not a Swede or a Belgian?) has a good and simple way of doing this it would certainly be a good thing. My idea of the danger section was to make it a somewhat rapid flutter of eclipses or flashes, say, flash and eclipse in period of two seconds or so. What do you think of this?

I wrote some time ago to the assistant secretary (Mr White) of Royal Society for your last year's proposal paper but have not heard in answer yet. I shall see Professor Stokes to-day I hope, and speak to him about it. Have you any changes or additions to be made to it?

Yours truly,
[Signed] William Thomson

The final paragraph refers to the nomination of Hopkinson for election to the Fellowship of the Royal Society. He was duly elected a few months later.

Consulting engineer and expert witness

When the move to London to set up as a Consultant was made in 1877 there were three children and the family set up house in Kensington not far from Addison Road – now Kensington (Olympia) – station. The first eighteen months of building up the consulting practice appear to have been somewhat anxious and straitened financially owing to losses resulting from a reversal in the fortunes of the Chatterley Iron Company in which the very considerable family savings had been invested.

A period of invention, development and rapid expansion like the eighties and nineties inevitably produces a spate of technical and scientific litigation, and the expert opinion of scientific men tends to be much in demand in the Law Courts. John Hopkinson soon found himself involved in this type of activity and he displayed an astonishing aptitude for it. Not only did he possess a scientific background sufficient to place him in the front rank, but he combined with it sound engineering experience.

He was capable of exceptional mental concentration and rapidity of thought and he was able to express himself, in court, with a brevity and lucidity that carried conviction. It is on record that, as an expert witness, the great Lord Kelvin sometimes cut a comparatively poor figure beside him. Expert evidence in patent actions soon became one of the most lucrative parts of Hopkinson's practice and so remained.

The great impetus to the development of the public supply of electricity came with the successful 'subdivision of the electric light', that is to say, the development of small electric light sources suitable for domestic application as distinct from the electric arc which had achieved great success in the lighting of public buildings and open spaces.

In December, 1878, Joseph Swan demonstrated to the Newcastle Chemical Society a carbon filament incandescent lamp in operation. Edison, in the United States, who had been working on the same problem, filed his patent application in November, 1879, to be followed by Swan in January, 1880. In 1881 the two inventors abandoned litigation and the two English companies merged as the Edison Swan United Electric Light Company. Hopkinson's entry into the field of consulting engineering thus coincided with the beginnings of electricity supply on a commercial scale. Within a very few years the first successful attempts were made to apply electricity to traction.

For the next twenty years Hopkinson's involvement in many of the major electrical engineering projects of the day resulted in his making fundamental contributions to dynamo design, to the theory of the operation of alternating current machines, and to electrical engineering practice. In parallel with this, he pursued his scientific researches, making advances in dielectric theory and in knowledge of the behaviour of dielectrics and of the magnetisation of iron which have taken their place in the classical literature of physical science.

Scientific research: Dielectrics

In 1878, at the age of 29, John Hopkinson was elected to the Fellowship of the Royal Society. The scientific work which was thus recognised was an extensive series of researches on the dielectric properties of glass and some other transparent media. Being engaged in the manufacture of lenses and prisms for lighthouse lanterns, it was natural that Hopkinson should be interested in the optical properties of glasses. A specific problem was, however, presented to him by Sir George Stokes, Professor of Natural Philosophy in Cambridge, in regard to possible compositions which would achromatise with one another and, although the result of the particular investigation was negative, the association with Stokes was for Hopkinson to be of continuing significance.

The most powerful intellectual stimulus to which Hopkinson's mind responded was almost certainly Maxwell's *Treatise on Electricity and Magnetism* which had appeared in 1873. To his intense interest in Maxwell's work, and to the personal influence of Kelvin, may be attributed the impetus to the pursuit of the long series of researches on dielectric phenomena which Hopkinson undertook.

In his paper of 1876, entitled *The Residual Charge of the Leyden Jar*, Hopkinson recorded the experimental observations upon which he based the principle of the superposition of dielectric absorption currents. In this paper he attributes the residual charge of the capacitor to that part of the polarisation of the dielectric which is time dependent. He points out that glass may be regarded as a variety of different silicates, some of which will approach the limiting polarisation corresponding to a given electric force, relatively quickly, whilst others will do so slowly, and that a mathematical model which would represent such a system would be a set of first-order linear differential equations. A solution of the form $E = \sum_o^n A_r\, e^{-\lambda_r t}$ could correspond, for example, to the observed behaviour when a Leyden Jar was first charged positively for a long period, discharged, charged for a short interval in the reverse direction, discharged and then insulated and the subsequent potential variation at its terminals recorded. The potential would first increase fairly rapidly to a maximum in the negative direction, then decrease to zero, pass relatively slowly to a positive maximum and finally die away. 'These results' he remarks, in a footnote, 'are closely analogous to those obtained by Boltzmann for torsion. From his formulae it follows that if a fibre of glass, is twisted for a long time in one direction, for a shorter time in the opposite direction, and is then released, the set of the fibre will for a time follow the last twist, will decrease and, finally, take the sign of the first twist.'

It was Maxwell who suggested to Hopkinson the parallelism between the after effects of mechanical strain, as they had been observed and analysed by Boltzmann, and those of residual charge. This simple but fundamental concept served as a basis for dielectric theory until Debye, in 1912, published his work on polar molecules.

It is a fair assumption that Hopkinson's imagination had been fired by Maxwell's deduction that light was an electro-magnetic phenomenon and that for a transparent substance the square of the refractive index should equal the permittivity. His Royal Society Paper of 1878 on the electrostatic capacity of glass opens with a statement of these considerations and records measurements on four optical glasses, the results of which 'by no means verify the theoretical result obtained by Professor Maxwell . . .'. From time-to-time during the next twenty years he returned to the pursuit of this elusive subject.

Writing in 1881 on the electrostatic capacity of liquids, he remarks 'A glance shows that while vegetable and animal oils do not agree with Maxwell's theory, the hydrocarbon oils do. It must, however, never be forgotten that the time of

13

disturbance in the actual optical experiment is many thousands of million times as short as in the fastest electrical experiment, even when the condenser is charged or discharged for only the 1/20,000 second'.

His researches on glass appear to have been pursued in his private laboratory, utilising materials of known composition with which the works could provide him; while his experiments on dynamo machines and on arc lights were made at the works.

The Alternator, the Dynamo and the Electric Light

In London, Hopkinson ceased to have direct responsibility for the execution of engineering projects. He brought to his consulting work a rare combination of qualities, mathematical ability, scientific insight, a clear appreciation of practical requirements and substantial practical engineering experience. He entered the field at a time when the opportunities for the exercise of such talents were almost unlimited and he entered it equipped as were few of his contemporaries.

4   Siemens Dynamo, believed to be the machine upon which Hopkinson carried out the tests reported in his paper *On Electric Lighting* of 1879

Starting rather slowly at first, Hopkinson's exceptional ability as an expert witness became recognised and within about two years, patent work had become by far the most lucrative part of his practice.

It was natural that, in the period following Hopkinson's service with Chance, he should publish work relating to lighthouses or stemming from his lighthouse experience. In 1879–1880, two papers *On Electric Lighting* appeared in the Proceedings of the Institution of Mechanical Engineers. The first deals with extensive tests of the behaviour of 'a Siemens medium-sized machine, that is, the machine which is advertised to produce a light of 6000 candles by an expenditure of $3\frac{1}{2}$ horse power.' The performance of the machine, a series dynamo, is represented by a characteristic curve plotting terminal voltage against current – the current being expressed in Webers. This simple device for representing and comparing the performance of series dynamos immediately proved its utility and passed rapidly into accepted practice. The second paper describes photometric measurements upon the electric arc and shows how the characteristic curve of the machine may be used to determine conditions for the maintenance of the arc.

Paris in 1881 was the scene of a great international exhibition, L'Exposition Internationale d'Électricité. Hopkinson's involvement is described by his wife:

I can remember [she wrote] how, in 1881, John was made one of the judges of the first Electrical Exhibition ever held in Paris. There we stayed at the Hotel Chatham with all the leading electricians of the time. In the hotel were staying also Sir William and Lady Thomson, Dr Werner Siemens from Berlin, Du Bois-Raymond, the great surgeon, Herr von Helmholz, Fletcher Moulton, and many others . . . I remember, at night, my husband and I would wander into the long galleries of the exhibition, lit up by the great globes of the Brush incandescent light of a hundred volt power, the Jablochkoff and many other new firms just coming on. The blaze of light was intoxicating.

Again, referring to Kelvin, presumably on the same occasion, although her memoirs do not make this quite clear: 'We were invited to listen to the Opera *Robert le Diable* just being given at the Grand Opera House and for the first time telephoned. Sir William sat glued to the receiver, every now and then turning around "John, the amazement of it. I can even hear the feet of the ballet girls dancing".' Graham Bell had invented the telephone only some six years before.

In 1883 Hopkinson delivered before the Institution of Civil Engineers a lecture entitled *Some Points in Electric Lighting*. The lecture reads to-day as an admirable popular account of the main phenomena of the electric circuit and of dynamo electric machinery, alternating, as well as a direct current. Amongst other things he deals with the parallel running of alternators. 'Now I know', he says, 'it is a common impression that alternate current machines cannot be worked together, and that it is almost a necessity to have one enormous machine to supply all the consumers drawing from one system of conductors.' He goes on to describe qualitatively the phenomena associated with running alternators in series and in parallel and shows that stable running can be achieved automatically by the parallel connection. He adds 'A little care only is required that the machine shall be thrown in when it has attained something like its proper velocity. A further corollary is that alternate currents with alternate-current machines as motors may theoretically be used for the transmission of power.' He concludes the lecture, which was illustrated by a range of experimental demonstrations, with some account of the properties of the carbon filament incandescent lamp, very recently brought to successful development. 'The building', he states, 'is, this evening, lighted by about 200 lamps, each giving sixteen candles light when 75 watts of power are developed in the lamp'.

In the following year, Hopkinson presented to the Institution of Electrical Engineers his paper on the *Theory of Alternating Currents, particularly in reference to Two Alternate Current Machines connected to the same circuit*. This paper gives a fundamental theoretical demonstration of the principles under-

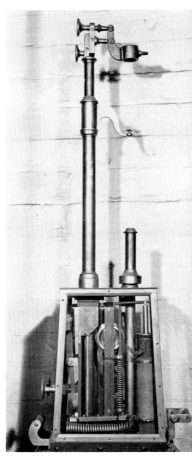

5   De Meritens Alternator installed at the Macquarie lighthouse

6   The Arc Mechanism of the Macquarie light

lying the parallel running of alternators. At the end of the paper he describes the practical verification of the theory with two of the De Meritens machines built for the electrical installation of the lighthouse at Tino. He also refers to the experiments carried out in co-operation with Professor Grylls Adams at the South Foreland. In a footnote, dated 22 November 1884, to the paper as printed in the Proceedings of the Institution of Electrical Engineers, Hopkinson states:

My attention has only today been called to a paper by Dr Wilde published by the Literary and Philosophical Society of Manchester, December 15th 1868, also Philosophical Magazine, January 1869. Dr Wilde fully describes observations of the synchronising control between two or more alternating current machines connected together. I am sorry I did not know of his observations when I lectured before the Institution of Civil Engineers, that I might have given him the honour which was his due. If his paper had been known to those who have lately been working to produce large alternate current machines, it would have saved them both labour and money.

This generous recognition of work which had anticipated his own was characteristic of the man.

It was not until 1886 that Hopkinson read to the Institution of Civil Engineers a paper describing the important features of the optical and electrical equipment of the lighthouses at Macquarie and Tino. At Tino the two de Meritens alternators were arranged for parallel operation. The Macquarie lighthouse, built in 1818 on the South Head near Sydney, was the first in Australia to be equipped with electric light.

Early in 1882 Hopkinson accepted the invitation of the recently formed English Edison Company to be one of their consulting engineers. It was in a report to the directors of this company that he expounded the principle of the two-part tariff method of charging for the supply of electricity. The company did not, however, pursue the project of setting up supply stations and some ten years were to elapse before the idea was publicised in his Presidential Address to the Junior Engineering Society in 1892 on *The Cost of Electricity Supply*. In this he remarks:

The ideal method of charge then is a fixed charge per quarter proportioned to the greatest rate of supply the consumer will ever take and a charge by meter for the actual consumption. Such a method I urged in 1883 and obtained the introduction into certain Provisional Orders of a clause sanctioning "a charge which is calculated partly by the quantity of energy contained in the supply and partly by a yearly or other rental depending upon the maximum strength of the current required to be supplied".

Dynamo design and Magnetic measurements

It was in the autumn of the same year that Hopkinson wrote to the company recommending a critical study of the Edison dynamo with a view to possible improvement and the modification of the design to meet varying requirements.

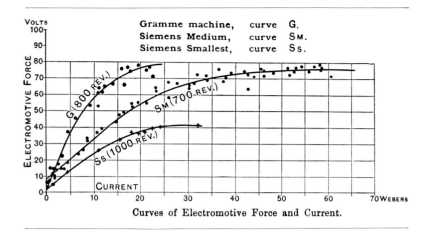

Curves of Electromotive Force and Current.

7 Characteristic curves of three dynamos, from Hopkinson's second paper *On Electric Lighting*

The manufacturing rights of the Edison dynamo were held, in this country, by Mather and Platt of Manchester; they were acquired from Edison following a visit by William Mather to the United States. The chief engineer of this company was John's younger brother, Edward. The outcome of this investigation is contained in two papers, one in the Philosophical Transactions of the Royal Society, 1886, and the other in the Proceedings in 1892, having the common title *Dynamo-Electric Machinery*. Pretty well at the outset he remarks:

> One purpose of the present investigation is to give an approximately complete construction of the characteristic curve of a dynamo of given form from the ordinary laws of electro-magnetism and the known properties of iron, and to compare the result of such construction with the actual characteristic of the machine. The laws of electro-magnetism needed are simply (1) that the line integral of magnetic force whether in iron, in air or in both, is equal to $4\pi nc$ where c is the current passing through the closed curve and n is the number of times it passes through; (2) the solenoidal condition for magnetic induction, that is, if the lines of force or of induction be supposed drawn, then the induction through any tube of induction is the same for every section.

From these fundamental relations is developed the principle of the magnetic circuit–a principle basic to the design of almost all electro-magnetic machinery and apparatus.

Clear priority and independence of conception are difficult to establish and are probably relatively rare. For the magnetic circuit, the question cannot be resolved for, in the minds of at least two of the leading electrical engineers, the concept was germinating at about the same time – Gisbert Kapp and John Hopkinson.

In November 1885, Kapp read to the Institution of Civil Engineers a paper entitled *Modern Continuous Current Dynamo Electric Machines and their Engines*, in which he remarked:

The author would here submit the formulas he is using in the designing of dynamos. In establishing them his aim has been to do away as far as possible with coefficients depending upon the type of machines and to put the expressions in a form sufficiently simple for practical use ... The strength of the field or number of lines Z is the ratio of the exciting power P to magnetic resistance R. In a dynamo machine the latter consists of three parts – R that of the air space which is considered to be independent of the strength of the field, and of the resistance of the armature core R_A, and of the field magnets R_F, both of which are considered to increase with the strength of the field:

$$Z = \frac{P}{R_a + R_A + R_F}$$

By the time Kapp's paper was read the Hopkinson modification of the Edison machine had been designed, built and tested and in the discussion on the paper Hopkinson outlined his design method. Basically, the principle was the same: Kapp, however, regarded each part of the magnetic circuit as having a permeability which, for the iron, was dependent upon the flux density, whereas Hop-

8 Edison Dynamo,
forerunner of the
Edison-Hopkinson machine

9 Experimental dynamo field models used by Hopkinson in the re-design of the
Edison machine

kinson worked directly from the magnetisation curve of the material. Hopkinson's analysis was the more elegant and its development the more thorough.

Mrs Hopkinson recalls, in the memoir *The Story of a Mid-Victorian Girl*, a Sunday morning discussion between John and Edward at the Royal Huts Inn at Haslemere, in which the plan for the redesign of the dynamo was laid down by the older man. The problem was tackled theoretically by developing the magnetic circuit concept and experimentally by the use of models of the old magnet system and of the proposed modified system. These models which were made by the instrument maker, William Groves of Bolsover Street, London, are preserved in the Science Museum at South Kensington.

On the theoretical side the magnetic circuit principle is applied to the approximate main dimensioning of the magnetic field system and the armature, approximate calculations of magnetic leakage are made and the characteristic surface of the machine derived. In addition, commutation conditions in the

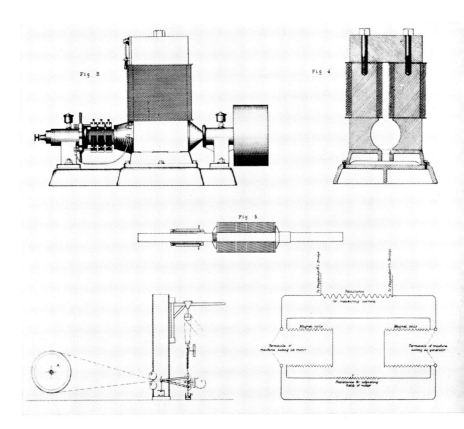

10 The Edison-Hopkinson machine, as illustrated in the Philosophical Transactions of the Royal Society, 1886, together with the dynamometer and the electrical circuit used in the 'back-to-back' testing of two identical machines

machine are considered. The modifications which were made appear absolutely obvious today. The Edison dynamo, with its long slender bipolar field, was clearly carrying a leakage flux which was a substantial fraction of the useful flux linking the armature. The modified machine had a short compact field system. The design of the machine, which was built by Mather and Platt, and the extensive tests on it, are detailed in the papers, together with corresponding data for the 'Manchester' dynamo which – in the different arrangement of its magnet system – appears to be related to the earlier Siemens dynamo. Of particular interest is the section entitled 'Efficiency Experiments', for here is disclosed the now commonly employed 'back-to-back' method of testing, on load, of two identical machines. It is often stated or implied, quite correctly, that the back-to-back method is employed, where practicable, in order to avoid the problem of providing and absorbing the large amounts of power which may be involved in testing electrical generators or motors on load. The test was not invented by

11   Two Edison-Hopkinson machines undergoing 'back-to-back' test

Hopkinson for this purpose. He was confronted with the difficulty of making an accurate *mechanical* measurement of the total power input or output of a machine. He explains that if only the *difference* in the power between two identical machines mechanically coupled, one motoring and the other generating, requires to be measured 'even a considerable error in the determination has but a small effect on the ultimate result'. The first machine of the new design gave, on test, more than double the output of the Edison dynamo of the same weight.

It was Hopkinson's interest in dynamo-electric machinery that directed his attention to more fundamental work on iron and ferromagnetic materials. His immediate requirement was for data on the magnetic materials used in the construction of dynamos and in pursuit of this information, he devised methods of measurement and made important fundamental observations. His paper *The Magnetisation of Iron* in the Philosophical Transactions in 1885 gives magnetisation curves for samples of wrought iron, steels with different heat treatments and cast iron, and in two tables are set out the magnetic properties, the specific resistance and the chemical analysis of the thirty-five specimens tested. The term *Coercive Force* is introduced as 'that reversed magnetic force which just suffices to reduce the induction to nothing after the material has been submitted temporarily to a very great magnetising force' and it is noted that 'for any practical purpose we may assume that the greatest dissipation of energy which can be caused by a complete reversal to and fro of magnetisation is approximately measured by:

$$\frac{\text{coercive force x maximum induction}}{\pi}$$

The method of measurement employed the 'bar and yoke' equipment which has served as a prototype for much of the apparatus subsequently employed in magnetic testing.

It appears that the instrument maker, Mr Groves, drew Hopkinson's attention to the non-magnetic property of Hadfield's 12% manganese steel and this is confirmed by data given in the table.

Four years later, Hopkinson reported to the Royal Society on the other work on magnetic materials for which he is best known – his investigations of the phenomenon of recalescence. The Curie temperature or magnetic change point at which iron ceases to be ferromagnetic is around 800 °C, a cherry-red heat, and a mass of iron cooling through this point glows momentarily more intensely as the change point is passed. Hopkinson measured the amount of heat evolved at the point at which iron returns from the non-magnetisable to the magnetisable state and further demonstrated that there is no pause in the cooling curve of non-magnetisable manganese steel. He was, it appears, the first to demonstrate

12   Manchester Dynamo. A machine with a different form of magnet system, compound wound, and rated at 105 volts, 130 amperes, at 1050 r.p.m.

the connection between recalescence and the disappearance of magnetism at the well defined temperature now known as the Curie point.

In his Birmingham period, as has been mentioned, Hopkinson's experimental work in fundamental science was carried on in the small laboratory which he had set up in two rooms of his house. While Mrs Hopkinson's account does not record that the same practice was followed when the family moved to London, it may safely be assumed that it was, for until 1890 Hopkinson had no other regular laboratory facilities at his disposal in London. Certainly when the family moved, after a very few years, from Holland Villas Road in Kensington to a fine spacious house with a large garden on Wimbledon Common, a laboratory was set up there. An entry in Hopkinson's diary for March 31st, 1896, records 'For some time now Jack and I have been experimenting on Röntgen rays, the laboratory having been moved downstairs.'

In 1890 he was invited to accept the post of Professor of Electrical Engineering in King's College, London. The conditions of appointment permitted him to carry on his consulting practice and required him only to direct the teaching and the work of the laboratories. He thus acquired physical resources and the help of staff and students for the pursuit of experimental researches. The name of Ernest Wilson, who joined him as demonstrator in 1891, appears as joint author along with F. Lydall, of a paper on *Magnetic Viscosity*. In the following year Wilson is again joint author of a paper entitled *Propagation of Magnetisation of Iron as affected by the Electric Currents in the Iron*, and acknowledgment is made for help with the experimental work to three student demonstrators, Brazil, Aitchison and Greenham.

The British Association for the Advancement of Science had, from its foundation in 1831, not only provided a platform for the reporting and discussion of scientific advances but had itself supported research in various ways. Typical of the association's research activity was the setting up, at the suggestion of Sir William Thomson, of a committee to consider the position of units and standards for electrical measurements, and in particular a standard of electrical resistance and its practical realisation. This committee which, in addition to Thomson, included in its membership Wheatstone, Bright, James Clerk Maxwell, C.W. Siemens, and Balfour Stuart, in its early reports first demonstrated how, following the work of Weber, the various electrical and magnetic quantities were related to the fundamental entities of length, mass and time. Definitions of electrical units deriving from these relationships then led to the problem of translating these definitions into practical standards of measurement which could be used in manufacture and in commerce. To this important and exacting task the committee devoted some eight years of work and created the first practical resistance standard – the B.A. ohm.

The sequel to the physical realisation of the unit of electrical resistance was, quite naturally, the problem of the degree of accuracy and the stability with which its realisation had been achieved and this question brought Hopkinson into membership of a B.A. Committee 'appointed for the purpose of constructing and issuing practical standards for use in electrical measurements'. His name appears on the numerous reports of this committee from 1882 until 1890. Much of the work of re-checking resistance standards was done in the Cavendish laboratory in Cambridge and measurements were made in other University and industrial laboratories but there is no evidence that Hopkinson himself undertook or directed any of the experimental work.

Hopkinson had joined the B.A. in 1870 and, apart from his work on the Electrical Standards Committee, he contributed a number of papers and he served as a member of the Council. In 1882 at Brighton his paper was entitled *On the stresses produced in an elastic solid by inequalities of temperature*, while in 1875 and 1886 he spoke on matters relating to lighthouse illumination.

Electric power and Electric traction

When Hopkinson set up his consulting practice in London, electric arc lighting had been established on a considerable scale in public places, railway yards, theatres and to some extent in factories. In 1880 Edison virtually initiated the public supply of electricity. He obtained a basic patent in the U.S.A. covering various components of a supply system – incandescent lamps, generators and auxiliary equipment. His system of supply incidentally envisaged the distinction between a feeder and a main, subdivision thus minimising to the required degree the voltage variations amongst lamps connected in parallel throughout the length of the main. He formed the Edison Electric Illuminating Company and established the Pearl Street Station in New York. In London, pioneer efforts at public supply followed rapidly. Private bills had already been presented in

13 Leeds Tramcar, powered by two motors having single reduction gear and operated on series – parallel control

25

Parliament to obtain Statutory Powers to permit the laying down of mains under public thoroughfares, a requirement already existing for gas and water undertakings. There was, of course, no specific legislation relating to electricity supply. Accordingly, the Government set up a Select Committee with Sir Lyon Playfair as Chairman to investigate the powers of public authorities in this matter, and to consider the conditions under which private companies should be permitted to provide a supply. Hopkinson gave evidence to this Committee.

Based on the Committee's report the first Electric Lighting Act was passed by Parliament in 1882. Providing, as it did, all the necessary facilities, the passing of the Act should have been followed rapidly by the establishment of undertakings financed by investors ready to exploit the potentialities of this new and promising service to the public. Unfortunately, one of its provisions proved to be too discouraging to investors – municipalities were given the right to acquire undertakings after twenty-one years. So repressive was the effect of this clause, that, six years later, an amending Act was passed increasing the period to forty-two years. The passing of this amending Act marked the beginning of the increasingly rapid growth of the electricity supply industry.

Meanwhile, attempts were being made to apply electricity to traction. Following the demonstration by Siemens and Halske of an experimental electric railway at the Industrial Exhibition in Berlin in 1879, a quarter mile length of electric railway was put in operation in Brighton in 1883 and shortly afterwards six miles of electric tramway went into service between Portrush and Bushmills in Northern Ireland. By 1891 the City of Leeds had a complete electric tramway system and the place of electricity in rail traction had been firmly established.

Hopkinson's advice was sought on many of the projected schemes for electricity supply or for electric traction. When the Metropolitan Electric Supply Company was formed in London in 1886, Hopkinson was appointed Consulting Engineer. A little later, 1891, he carried out the design of the Manchester Electric Lighting Works. The electric tramway from Kirkstall to Roundhay in Leeds was constructed under his supervision, as were also the first electric lines in Liverpool.

The Manchester Electric Lighting Works were described by Hopkinson at a meeting of the Institution of Mechanical Engineers in July 1894. The initial installation comprised four main bipolar generators producing 600 amperes at 400 volts. Distribution was by five parallel conductors, middle conductor earthed providing four 100-volt circuits, balance being maintained by four small 100-volt machines. The arrangement was a modification of the three-wire system which Hopkinson had invented in 1881. Flat bare copper conductors were employed, carried on insulators, and running within concrete ducts. Connected to the mains were 18,600 incandescent lamps of 16 candle power, together with 250 arc lamps and motors totalling 16 horse power. It was in the Manchester electricity supply system that Hopkinson's proposal for a two-part supply tariff was first introduced. The average price of electricity to the consumer was 5½d per unit.

Distributing by bare conductors under the damp conditions existing in the ducts, produced a phenomenon which became familiar to d.c. distribution engineers and which, in his paper, Hopkinson described and explained. There

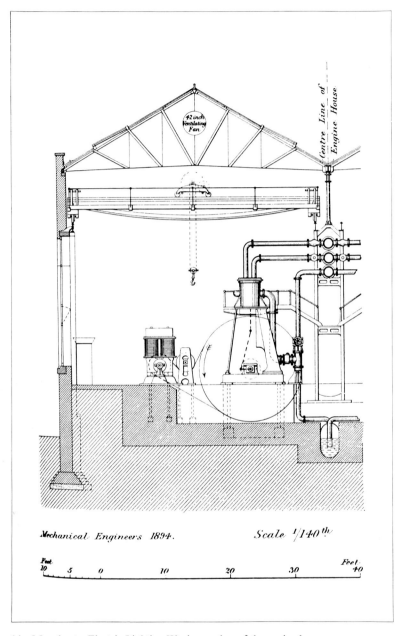

Labels within the figure:

42 inch Ventilating Fan

Centre Line of Engine House

Mechanical Engineers 1894.

Scale 1/140th

Feet 10 5 0 10 20 30 Feet 40

14 Manchester Electric Lighting Works, section of the engine house

was a permanent small leak on the whole system, but the insulation resistance of the positive conductor was found always to be much higher than that of the negative. This was due to osmosis, which removed moisture from the region of the insulators of the positive conductor and deposited it in the region of the negative.

Reference was made in the discussion on the paper to the use of ball bearings in the small generators. Hopkinson commented that 'ball bearings which he had in use at his own house, with a small dynamo machine, had been working for nearly a year with perfect satisfaction, the machine running at 1400 revolutions per minute'. The best testimony to their efficiency was that his stable boy, who had charge of the machine, had never let him hear anything about them.

On other projects, Hopkinson served in the capacity of consultant to one of the main contractors. Probably his most noteworthy service in this capacity was as consultant to Mather and Platt in relation to the City and South London Railway – the first tube railway in the world, and indeed the first true electric railway in the sense of having a line with defined stations and a proper system of signalling. Mather and Platt were contractors for the generating machinery and rolling stock and were under obligation to run the system for two years. Clearly, in this project, John co-operated closely with his brother, Edward, who as has been mentioned, was Chief Engineer of the Company. Edward Hopkinson had the advantage of previous experience of electric traction as Consulting Engineer for the Bessbrook & Newry tramway in Ireland, and earlier of the Portrush and Bushmills line, while he was on the staff of Siemens Brothers.

15   The Bessbrook & Newry tramway showing the bow current collector

Edward Hopkinson published in the Proceedings of the Institution of Civil Engineers in 1888 a paper on the Bessbrook & Newry tramway and in this he refers to John having devised, for the project, a bow type of current collector to be used with an overhead conductor at an oblique road crossing at which the conductor rail and shoe would be impractical.

At the Stockwell generating station there were initially three, later four, large Edison-Hopkinson compound dynamos with an output of 450 amperes at 500 volts. Trains consisted of three cars drawn by a locomotive weighing 10.3 tons. Current was collected from the off-centre conductor rail through flat, cast-iron shoes, and the drive was furnished by two series-wound bipolar motors with Gramme ring armatures mounted directly on the axles, the pole-piece ends of the field magnets being carried by bearings on the axle. The direct axle mounting of the armature was adopted in order to avoid the use

16 City and South London Railway, generating station at Stockwell

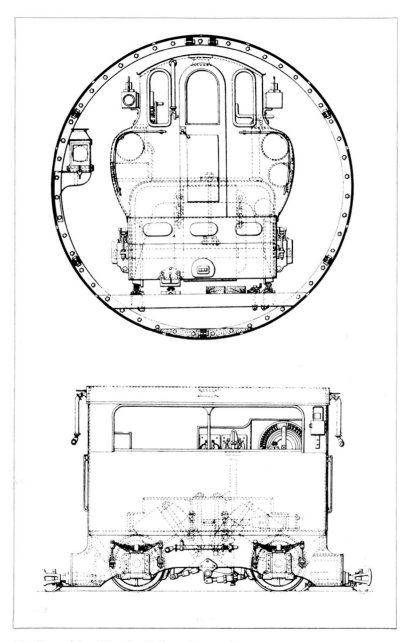

17   City and South London Railway. Locomotive

of gearing, which, it was found, would give rise to excessive noise in the tunnels. Their combined maximum output was 100 h.p. developed at a motor speed of 310 r.p.m. and a train speed of 25 miles per hour. The first fourteen locomotives were built by Beyer Peacock and equipped by Mather & Platt. The system started operating in 1890 and it is interesting to record that locomotives generally of the type originally installed continued in service on the City and South London tube until 1923. The performance of the trains individually proved to be satisfactory but difficulties arose when ten or more were running at the same time. Although, quite evidently, the whole concept of the system was sound and had set the pattern for future development, it remained under some handicap of various technical limitations and, in its own right, never really achieved commercial success.

It is probable that a pioneer development such as this was the type of engineering responsibility in which Hopkinson found the greatest satisfaction. In his memoir, his son Bertram writes:

it may be doubted whether Hopkinson would have found much interest in problems which had already been pretty fully worked out by others. In constructive engineering he liked to do one big job of a kind but did not care about the work of the same sort which would necessarily follow its successful accomplishment. A large practice in traction or lighting engineering had no attraction for him. He had one great desire and that was to carry out a long distance transmission of power . . . It was a dream the realisation of which was probably only prevented by his death.

It is however on record that he declined an invitation to enter the competition for a scheme for the generation of electric power from the Niagara Falls which was announced in 1890. He replied a trifle curtly that he did not devote time to engaging in competitions.

Patents and Patent actions

It would, in one sense, be true to say that during the whole of his life in London, Hopkinson's major activity was his work as an expert witness in patent actions, for, as has been mentioned, by far the greater part of his income came from his work in the Law Courts. In this work, Hopkinson set for himself the highest standards of integrity, application and professional competence. None the less, he appears to have handled scientific-legal problems with extraordinary ease and rapidity. His success may well have been due, in part, to the zest with which he would engage in a battle of wits when under cross-examination.

Several of the more important of the earlier actions in which he was engaged concerned telephone communication and apparatus. One such case was that of the Attorney General *v.* The Edison Telephone Company of London, reported in the Law Reports of 1880, in which it was held that the Postmaster-General's exclusive right to transmit telegraph messages, conferred on him by the Telegraph Act of 1869, also embraced telephone messages, although the telephone had not been invented in 1869. Hopkinson, for the Edison Company (the Defendants) argued unsuccessfully that the Company did not itself, transmit messages but merely provided the *means* whereby the two subscribers personally transmitted their messages!

Quite a number of actions concerned alleged infringements of the Edison and Swan carbon filament incandescent lamp patents, Hopkinson giving evidence for the Plaintiffs.

Another of the more important actions in which Hopkinson was involved was Ferranti's petition for the revocation of Gaulard and Gibbs transformer patent, reported in 1889 and on appeal in 1890. The famous electric lighting installation at the Grosvenor Art Gallery in New Bond Street utilised, initially, the system of transformers in series invented by Gaulard and Gibbs. Ferranti introduced the parallel connection of transformers shortly after becoming Chief Engineer of the Grosvenor Gallery in 1886. Judgment was given in favour of Ferranti, and this was upheld on appeal and subsequently in the House of Lords, although the grounds for the decision varied somewhat. Hopkinson gave evidence on behalf of the Patentees, but his evidence was confined to construing the prior art and giving his views thereon. He must have been fully aware of the weakness of his client's case. This was probably an instance of the exercise of that quality which his son Bertram regarded as supremely important and of which he wrote: 'But perhaps his capacity of drawing the line between fact and argument was that which most distinguished him . . . Hopkinson, alike in the witness-box and in the laboratory, knew

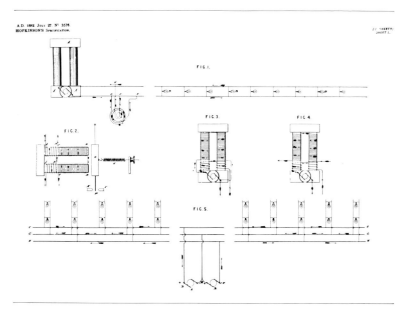

18   Drawings of Hopkinson's '3-wire' patent specification

exactly how far he was bound by the facts and where difference of opinion could come in'.

Unquestionably, Hopkinson was one of the foremost expert witnesses of his day, but it is interesting to record a comment made some fifty years later by Sir James Swinburne, another distinguished electrical engineer, who knew him and who had considerable professional contact with him. Swinburne wrote: 'Hopkinson was a little apt to be dead certain, and not always right, though he was very seldom wrong. I had a good deal to do with him in patent litigation and, when on the same side, sometimes found he had jumped to unsound conclusions'.

Hopkinson was himself the patentee of over forty inventions, most of which were made between 1879 and 1889, and of these several turned out to be of major importance. The most successful were the three-wire system of distribution and the series-parallel connection of traction motors. These represent the ingenious direct application of exceedingly simple principles. Patent No. 3576 of 1882 covers the three-wire principle in the following terms:

For the purpose of economising the cost of main conductors I place two dynamo machines in series and place two systems of lamps or other appliances consuming electricity of approximately equal capacity also in series. A main conductor is taken from each extreme pole of the two dynamos to points between the two systems of lamps, the intermediate conductor serving to bring back to the central station any electricity required for one system of lamps in excess of that required by the other system of lamps.

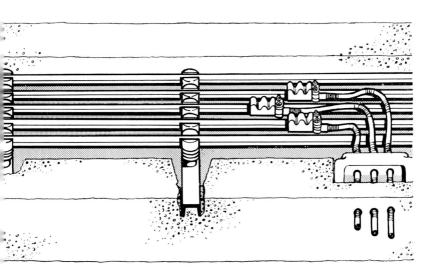

19 Reconstruction of Hopkinson's five wire distribution system

This patent, which had been sold to the American Westinghouse Company, became the subject of one of the most notable actions in which Hopkinson was involved. The case was Hopkinson *v.* The St James' and Pall Mall Electric Light Co Ltd, reported in 1893, infringement by the St James' and Pall Mall Electric Light Company being alleged. The action, which engaged the highest talents of Bench and Bar, lasted for three weeks between November 1892 and January 1893. It was tried before Mr Justice Romer in the Chancery Division of the High Court. Mr Fletcher Moulton, QC, and Sir Richard Webster, QC, appeared for the Plaintiff; Sir Horace Davey, QC, and Mr Finlay, QC, appeared for the Defendants. Lord Kelvin and Hopkinson himself were among the expert witnesses for the Plaintiff. Professors Silvanus Thompson and Kennedy gave evidence for the Defendants. Mrs Hopkinson records how she and Lady Kelvin attended day after day and sat fascinated by what she describes as 'the legal gladiatoral display'. Fletcher Moulton asked her one day at lunch if she did not think it as good as a play. The validity of the three-wire patent was upheld.

The characteristics of the series-wound direct current motor proved to be admirably adapted to the needs of electric traction. Starting the electrically-driven vehicle from rest or controlling the power of the motor was effected very simply by inserting resistance in series with the machine by means of a multi-contact controller providing a sequence of suitable 'steps' of resistance. Hopkinson's Patent No. 2989 of 1881 proposed that:

> In order to economically vary the speed of a tramcar, tram engine or other vehicle driven by electricity, two dynamo electric machines may be used on the car engine or vehicle and when high speed is desired these are arranged in parallel circuit; when a low speed and greater tractive force is required they are arranged in series.

By this means less power was dissipated in resistance during control operations. The advantages of series parallel control were immediately appreciated and the system almost universally adopted.

The majority of Hopkinson's patents did not result in successful commercial exploitation. Some of the inventions upon which he spent most time and trouble, for example, an ampere-hour electricity meter, were not practically successful, while others, like the three-wire system, paid great dividends for little labour.

University teaching and research

John Hopkinson had a natural aptitude and a liking for the imparting of information. He helped his younger brothers with their studies and, later, his own children. The pleasure and satisfaction which he derived from watching another mind grasping, with his help, some new concept is described by his son. He was interested in the broader aspects of education and the development of the Engineering School in Cambridge was a matter of close personal concern to him. He declined the invitation of the University to follow James Stuart in the Chair of Engineering, but he was instrumental in securing the appointment of Ewing to the office, an appointment which was of great and lasting benefit to the School. In due course, Hopkinson's son Bertram followed Ewing in the Chair and filled the post with distinction. It is interesting to read, in a memorandum on engineering education written by Hopkinson for a Cambridge University Syndicate in 1890, his opinion that:

> the majority of engineers in the country must be practical men who can deal effectively with the actual execution of a limited class of work, and for these it is probably best that their workshop education should begin earlier than the usual time of leaving the University. 'But. [he went on] there is also room in the profession for others who have a wider education addressed to enabling them to deal with circumstances for which there is no direct precedent. For this, I believe that the mathematical education of the University would be of great value; but with it should be coupled constant contact with the physical phenomena in some shape or other, so that the men may realise that every bit of their mathematical knowledge has its counterpart in objective facts. Of course, for such men the more nearly their physical work bears upon practical engineering the better.

20 Professor Hopkinson, Professor Wilson and Senior Students at King's College, London, April 1894. Back row: Jones, Aitcheson, Adlard. Front row: J. S. Highfield, John Hopkinson, Ernest Wilson and another

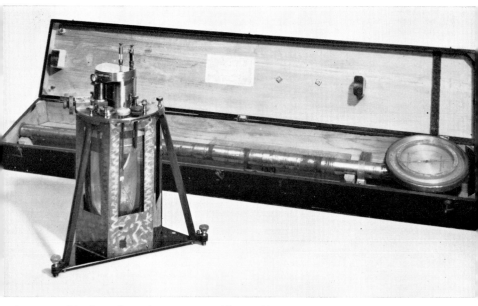

21 A quadrant electrometer used by Hopkinson in dielectric measurements, together with his hot-wire voltmeter in its carrying case

A graduate of the University of London of nearly twenty years' standing, Hopkinson was in 1887 appointed a Fellow of the University, thereby becoming a member of the Senate. The University of London was at that time purely an examining body and Hopkinson held a firm conviction that the system of External Examinations with its wide availability was a unique educational service of the highest value to the community.

King's College had received in 1890, from the widow of Sir William Siemens, a benefaction for the establishment of a department of Electrical Engineering within the department of Engineering, and Hopkinson was invited to accept appointment to the new Chair of Electrical Engineering. As has been mentioned, the conditions of appointment required him only to direct the teaching and the work of the laboratories. Hopkinson's appointment was of great benefit alike to himself and to the College. He now had staff and the equipment to sustain his researches and undergraduate students could themselves take part in suitable investigations. It appears that Hopkinson did little lecturing but taught by personal contact with students in the laboratory and in direct discussion. From time to time students would be invited to his home in Wimbledon. He aimed to develop the student's power to solve problems on his own resources and he brought to the College from his consulting practice a sequence of current problems upon which his men could begin to build up their experience. For more than a generation after his death, Hopkinson's students continued to make their mark in the engineering world.

The Proceedings of the Royal Society, the Journal of the Institution of Electrical Engineers, and the pages of *The Electrician*, bear witness to the fruitful association of John Hopkinson with King's College. A few of his researches were published under his name alone, but most were joint publications with student demonstrators who assisted him, and with Ernest Wilson who, later, succeeded him in the Chair.

The Engineering Institutions

Hopkinson was elected a member of the Institution of Mechanical Engineers in 1874 and of the Institution of Civil Engineers in 1877, and several of his papers were read to one or other of these Institutions. He served on the Councils of both. He was, however, also active in promoting the interests of the then very young sister Institution, the Society of Telegraph Engineers, which, in 1889, became the Institution of Electrical Engineers. He was elected a member in 1881. Hopkinson enjoyed the rather rare distinction of serving two terms as President of the Institution of Electrical Engineers, first in 1890 and again in 1896. In 1890 his address was on *Magnetism*, and in 1896 he reviewed the state of knowledge of electrical science and examined critically the systematic arrangement of the subject as presented to electrical engineers. Referring in his address to Maxwell's theory, he remarks, 'Whether the postulate of an all pervading ether be, or be not, a metaphysical necessity, surely it is well for the practical man and the physicist to leave the question to the metaphysician'. Hopkinson deprecated the introduction of assumptions which he regarded as artificial and unnecessary in the description of observed facts. He could, however, speculate upon occasion. Hopkinson was a member of the Royal Institution and on 26 April 1895, he gave one of the famous Friday Evening Discourses. Sir Frederick Bramwell was in the Chair and the title of the discourse was *The effects of Electric Currents in Iron on its Magnetisation'*. It was, in fact, a popular account of some of the magnetic researches which he was conducting in King's College at the time and, in the great Royal Institution tradition, admirably illustrated by experiments. He brought his lecture to a close by remarking:

> In conclusion, let us indulge in a little wild speculation, not because it is probable that it is in any sense true but because it is interesting. Suppose a magnet were made exactly like the one upon which we experimented but of the size of the earth and that some mighty electrician generated such a current in its copper coils as would give a magnetising force of 2.5 and then reversed it, it would take some thousands of millions of years before the rate of disturbance at the centre attained its maximum value. The speculation I suggest is this, is it not conceivable that the magnetism of the earth may be due to currents in its material sustained by its changing induction but slowly dying away?

When Hopkinson assumed office for the second time the clouds were gathering over the international scene for the storm of the South African war. He opened his Presidential Address, to the surprise of the members, by an appeal for military service. In the minutes of the meeting of the Council on 16 January 1896, appears the entry:

> The President stated that, it having occurred to him as very desirable that some steps should be taken by the Institution to render the technical knowledge of electrical engineers available for the defence of the country, he had communicated his views to Lord Kelvin and other members of the Council who were unable to be present that evening, all of whom had expressed themselves strongly in favour of the proposed movement, and he accordingly moved: 'that steps should be taken to render available for purposes of national defence the technical skill of electrical engineers and that a committee be appointed with instructions to take such steps as it thinks fit to attain that end'.

That resolution resulted in the formation of a volunteer corps of electrical engineers.

In his diary, on 7 January, about a week prior to his I.E.E. Presidential Address, Hopkinson records that he saw Sir Arthur Halliburton at the War Office and afterwards he and his brother Alfred, called on Lord Wolseley, who 'received us very encouragingly and finally told me to get a scheme drawn up of what could be done and suggested my conferring with Webber'.

With the authority of the War Office, the Electrical Engineers, Royal Engineers Volunteer London Division, was founded on 27 January 1897. Lord Kelvin consented to become Honorary Colonel while Hopkinson, with the rank of Major, took command of the Corps. Hopkinson organised the Corps to include not only members of the I.E.E. and the University of London, but qualified Electrical Engineers in various branches of the Industry. The work of the

22 Hopkinson's diary for 16th January, 1896. (When he proposed the formation of the Electrical Engineer Volunteer Corps)

Corps was concerned for the most part with searchlight equipment and with the laying and maintenance of electrically exploded mines used mainly in connection with coastal defence. His diary contains fairly detailed information as to the operation of this equipment. In the minutes of the Council of King's College there appears an entry: 'A letter was read from Professor Hopkinson suggesting that it might be advantageous to the College if the submarine mining electrical testing apparatus of the Electrical Engineers Volunteer Corps was deposited at the Siemens laboratory and the members of the Corps given access to it under proper conditions'.

Hopkinson's command was cut short by his tragic death within little more than a year and he was succeeded by Major – later Colonel – R.E. Crompton under whom the Unit gave distinguished service in the Boer War.

The sub title, *London Electrical Engineers*, survived in various Army units deriving from the original foundation until in 1967 all Territorial units were disbanded.

Although Hopkinson played a leading part in the work and development of the Institution of Electrical Engineers from 1881 onwards, he was not a very frequent contributor to discussions on papers. One interesting example is his comment on Mordey's paper on *Alternate Current Working*, read in 1889 and referring to the paralleling of alternators:

> To obtain a great control of one machine upon another it is not of itself desirable to have any large self-induction as Messrs Kapp and Forbes appear to think, nor is it desirable to have it as small as possible, as Mr Mordey appears to think when he says: 'if it (self-induction) were absent probably the machines would run parallel all right'. The machines will best control each other when $\dfrac{2\pi\gamma}{\tau}$, γ being the self-induction, is equal to the resistance of the armature circuit and the leads to the junction with the leads of the other machine.

Mountaineer

Hopkinson was at the height of his powers when his life was cut short by a climbing accident in Switzerland. He was an experienced mountaineer and his enthusiasm for climbing was shared with the children who often accompanied him. On the morning of 27 August, Hopkinson, with his son Jack and his daughters, Alice and Lena, set off to make an ascent of Petite Dent de Veisivi by way of the Col de Zarmine. The party did not return and the next morning the four bodies were found roped together some seven hundred feet below the summit.

Evelyn Hopkinson met this overwhelming disaster with amazing courage. Only a few days before his death, Hopkinson had promised Ewing to open a subscription list for a proposed extension to the Cambridge engineering laboratories. On 13 October Hopkinson's widow wrote to the Vice Chancellor offering jointly with her son Bertram and her daughter Ellen, to give £5,000 towards the project as a memorial to her husband and her son Jack.

Climbing was, it appears, very much in the Hopkinson blood. All five brothers were members of the Alpine Club and they were pioneers of the development of rock climbing in Britain. Their names are commemorated in the Lake District by the Professor's Chimney in Scafell, Hopkinson's Cairn above Deep Ghyll and Hopkinson's Chimney on Dow Crags. The first recorded climbs on the North face of Ben Nevis are also credited to the Hopkinsons.

John Hopkinson was elected to the Alpine Club in 1889, his proposer being Sir Felix Schuster, the banker, and his seconder Sir Frederick Pollock, the distinguished lawyer. His first 'qualifying' climb is given in his application as being in the Alps in 1871. For Hopkinson no other physical activity presented attractions comparable with those of climbing, and his son Bertram describes vividly the zest and ebullient enthusiasm which possessed his father when engaged in some Alpine ascent.

A portion of his diary for the period 1895 to 1898 contains entries covering two family holidays in Switzerland which occupied rather more than the full month of August. Their climbing activities are described in detail – the composition of the party, the route followed, the weather, snow, ice and rock conditions, and the times to various identifiable points. There are comments on the travel arrangements, the abilities of the guides, and – with critical appreciation – the hotels and the food. These clearly are the notes of an enthusiast, recording the discriminating enjoyment of the connoisseur.

The record of John Hopkinson's life is of practically unbroken success in his public activities and in his private life, of domestic happiness. Perhaps to the outside observer, a man to whom success came so consistently, might be expected to be insensitive and impervious to external influences. His sister Mary, however, wrote of him: 'It troubled him to disappoint a child'.

Bertram Hopkinson, writing some three years after his father's death, the memoir which introduces the two volumes of his collected papers, remarks 'He was the most reserved of men and only his family knew him intimately'. He could be laconic in speech and occasionally, in correspondence, brief almost to the point of being curt.

As has been mentioned, some of Hopkinson's letters and a small part of his diary have been preserved, and in them there are pointers which enable one to penetrate a little way behind the front which he presented to the world. At the height of his career, his diary provides the occasional rather revealing passage. One of the 1895 entries records:

> Went to Stafford, driving to Addison Road. On to Manchester in the evening. Mother and Mary the only ones at dinner. Had much chaff with mother concerning morals. Father came in later.

Evidently his mother, for whom he had the deepest regard, must have accepted his iconoclastic badinage with tolerant good humour. The deeply religious parents regarded the theatre with, at worst, disapproval and, at best, suspicion. John and Evelyn, however, were fairly regular theatre goers and one entry relating to a theatre party reads:

> 16th March. Took a party of eight to see *King Arthur* at the Lyceum. One of the best tragedies I have seen. Ellen Terry, the great feature. Returned home

in a big covered wagonette, having supper as we went. We had the middle of front row of dress circle at the theatre. Dr and Mrs Thorpe were there.

The play *King Arthur* would probably be described nowadays as a spectacular and highly emotional!

By far the longest entries in the diary are those relating to climbing in Switzerland in the vacations. In one entry we read: 'I had my boots nailed with Swiss nails made at Neufchatel, which stood excellently'. Evidently every circumstance and detail of every climb was of significance – each experience something to be savoured in recollection.

There is also a long account of a family cycling holiday in Northern Italy. Due note is taken of scenic beauty, architecture, art and history, but one feels that perhaps just cycling was the most important thing!

That Hopkinson had a feeling for nature and for wild life is evidenced by many entries in the diary. For example, he records:

13th April. Walked with whole party to Fingle Bridge and then to Wifford Barton with Jack, saw many kestrels, rock doves, also swallows for the first time this year. Wild daffodils, primroses, celandine, wood anemones. Gorse and broom much destroyed by frost. Two old water mills in ruins.

In his garden at Wimbledon he followed the cycle of the seasons with close attention. Typical of many such entries is one for Thursday, March 18th, 1897, which reads:

Asparagus just coming up; just beginning in 1896 on March 25th. Early peas one inch high. Outdoor peaches pushing, some in bloom. Last year were coming into bloom on 21st. Gooseberries and currants in leaf; no bloom. Apples and pears pushing; not in bloom. Chestnuts just beginning to shew leaves rarely. Almond blossom not out. Vines breaking freely. On the whole things are about a week later than last year – indoor peaches a week earlier.

There are amongst the rather rare personal entries, two which are noteworthy. On January 10th, 1896, when, under the threat of war in South Africa, he was contemplating the proposal to found a volunteer corps of engineers, he wrote:

Am too much interested in the foreign crisis to do much effective work.

Probably he could more accurately have said that political events so weighed upon his mind.

Then on 21 February 1897 appears the last of such entries, which reads:

Have not been sleeping well and am getting into a generally lazy habit. As an experiment for one month – will not smoke at all and restrict myself to three glasses of wine a day.

On the whole the diary does not read like that of a man given, ever, to committing his innermost thoughts to paper. There is nothing to indicate that he was wrestling with scientific problems, that his resources – intellectual or physical – were being strained or that he was anxious about the outcome of some of the undertakings upon which he was engaged. One would guess that his barrier of reserve was let down only rarely even within his own family circle.

From the diary it appears that Hopkinson spent about a month on a visit to the United States in the autumn of 1895 but unfortunately the record of the trip was made separately and has not been traced.

It is doubtful whether Hopkinson had any intimate friends outside his family circle, but it is clear that he was not unsociable. The Hopkinsons entertained and visited a great deal and were on terms of friendship with many of the distinguished people of the time. Mrs Hopkinson describes in a most lively way visits to their home in Wimbledon by Sir William Thomson, Sir Benjamin Baker, Sir Edward Thorpe and Sir Henry Tate. Within his family Hopkinson must have shown something of that withdrawal and remoteness which go with daily concentration on exacting problems, but there was no barrier between him and his children. His daughter Ellen, who became Lady Ewing, wrote

23   Silver model of a searchlight presented to Mrs John Hopkinson in 1901 by the Electrical Engineers R. E. Volunteer Corps

'without a trace of sentimentality, he had much tender toleration of his children's foibles which he sought, unconsciously perhaps, to eradicate by example rather than by precept.'

Hopkinson cared little for honours, but two that gave him special pleasure were the award of the Royal Medal of the Royal Society for his researches on the magnetic properties of iron and his election to the Atheneum in 1887 as one of the nine annually chosen for eminence in literature, art or science.

There was a good deal of the fighting instinct in Hopkinson's make-up and something of an element of daring. Possibly his legal work satisfied to some degree his fighting spirit, but climbing provided that element of risk and physical danger without which satisfaction could not be complete. He possessed a rare combination of qualities, mathematical ability, scientific insight and a clear appreciation of practical requirements. Above all he possessed unswerving integrity and tenacity of purpose. He was trusted implicitly. His philosophy is perhaps well summarised in the concluding words of his lecture on *The Relation of Mathematics to Engineering*:

> whilst reverencing the discoverers who have added to our knowledge, we endeavour to replace their methods by better, and expect that those who come after us will, in their time, improve on ours. Our knowledge must always be limited but the knowable is limitless. The greater the sphere of our knowledge, the greater the surface of contact with our infinite ignorance.

His engineering achievements, which were characterised by the essential simplicity and directness of his application of scientific principles, would alone merit the honoured place which his name occupies in the annals of engineering and, with the advances which he made in pure science, place him in the first rank.

John Hopkinson was a man of great stature.

FURTHER READING

A memoir of John Hopkinson, written by his son Bertram, is included with the *Original Papers* of John Hopkinson, published in two volumes by the Cambridge University Press in 1901. The author has consulted memoirs printed for private circulation, in particular *The Story of a Mid-Victorian Girl*, Evelyn Hopkinson, Cambridge University Press, 1928; *John and Alice Hopkinson, 1824–1910*, Edited by Mary Hopkinson and Lady Ewing, with a preface by Sir Gerald Hurst, Farmer & Sons Ltd., London.

Readers of this booklet may also be interested in the Science Museum booklets *S.Z. de Ferranti* by Arthur Rıdding, published in 1964 and *R.E.B. Crompton* by Brian Bowers, published in 1969.

SOURCES OF ILLUSTRATIONS

3 *and* 7 Hopkinson: *Original Papers*

5 *and* 6 Museum of Applied Arts and Sciences, Sydney, Australia.

10 *Philosophical Transactions*, 1886

11 *and* 15 Mather & Platt Ltd

14 *Proceedings of the Institution of Mechanical Engineers*, 1894

16 *Proceedings of the Institution of Civil Engineers*, Vol. 73

18 *Forerunners of The North Western Electricity Board* by W.E. Swale, published by The North Western Electricity Board, 1963

Thanks are due to Mr. J.L. Willis, Director, and Mr. H.H.G. McKern, Deputy Director, of the Museum of Applied Arts and Sciences, Sydney, for the provision of photographs of the Macquarie lighthouse electrical equipment (plates 5 and 6).

Acknowledgement is also made of the kindness of Mr. and Mrs. Norman Robb of Princeton, British Columbia, in permitting the inclusion of the photograph of the silver model searchlight (plate 23).

Printed in England for Her Majesty's Stationery Office by Eyre & Spottiswoode Limited at Grosvenor Press Portsmouth

Dd 501658 K.60